# YOUR KNOWLEDGE HAS VALUE

Nadine Hänsel

# How and why transnational companies internationalize

GRIN Verlag

**Bibliografische Information der Deutschen Nationalbibliothek:**

Die Deutsche Bibliothek verzeichnet diese Publikation in der Deutschen National-
bibliografie; detaillierte bibliografische Daten sind im Internet über http://dnb.d-
nb.de/ abrufbar.

**Imprint:**

Copyright © 2012 GRIN Verlag GmbH
Druck und Bindung: Books on Demand GmbH, Norderstedt Germany
ISBN: 978-3-656-39227-9

**This book at GRIN:**

http://www.grin.com/en/e-book/211038/how-and-why-transnational-companies-
internationalize

**GRIN - Your knowledge has value**

Der GRIN Verlag publiziert seit 1998 wissenschaftliche Arbeiten von Studenten, Hochschullehrern und anderen Akademikern als eBook und gedrucktes Buch. Die Verlagswebsite www.grin.com ist die ideale Plattform zur Veröffentlichung von Hausarbeiten, Abschlussarbeiten, wissenschaftlichen Aufsätzen, Dissertationen und Fachbüchern.

**Visit us on the internet:**

http://www.grin.com/

http://www.facebook.com/grincom

http://www.twitter.com/grin_com

Christian-Albrechts-University of Kiel
*Globalisierung und regionalwirtschaftliche Entwicklung*
Summer Term 2012

01.08.2012

# How and why transnational companies internationalize

Nadine Hänsel

# Table of Contents

1. Introduction

2. How firms internationalize

   2.1 The Product Life Cycle
   2.2 Different Ways of TNC progress
   2.3 Born Global Firms

3. Why firms internationalize

   3.1 Market Access
   3.2 Lower Production Costs
   3.3 (Natural) Resources
   3.4 Transnational Strategy

4. Case Study: *Volkswagen*

5. Conclusion

6. Bibliography

   6.1 Research Literature
   6.2 Online Sources
   6.3 List of Figures

# 1. Introduction

*"I hear people say we have to stop and debate globalization. You might as well debate whether au-
tumn should follow summer. In the era of rapid globalization, there is no mystery about what works –
an open, liberal economy, prepared constantly to change to remain competitive."*
*(Tony Blair, see Lule, p. 25)*

This quote adequately summarizes that globalization is a temporal phenomenon,
which is irreversible and which therefore needs to be seen as an accomplished fact.
(Lule: 2012) With regard to the business world, increased competition forces companies to identify new markets and to cut costs while at the same time maintaining profit
by forcing down prices through the standardization of their products and the production process itself. International competition and international trade are no new phe-
nomena but the process of globalization intensified and accelerated these appear-
ances. (Hamilton; Webster: 2012)

The proper handling of competitors is certainly not the only reason for going international. The aim of this paper is to categorize other reasons of internationalization and
to identify possible ways of going international. The aim of this paper is to answer the
following questions:

1. Is there a best possible way for internationalization?

2. What economic incentives play a role when a firm decides to go international?

The focus of this research paper will be on transnational companies (TNCs), which
are not just doing business across borders, but rather have the individual characteristic of **direct** production and to a large extent also **direct** business activity in a foreign
country. (Gillies: 2012) What strategies TNCs choose in order to pursue their target
of directness will be demonstrated in this paper, which is roughly divided into three
parts.

The first part of the paper will be based on a look at a traditional view on the internationalization process, which will help to give a first understanding of the sequential
development a firm could pass through on its way of becoming international before
diverse trajectories will serve to allow deeper insights into the process of internationalization. In this context, the paper also offers an inquiring look at the so-called *born
global firms*. The second part will reveal several reasons, which express the motiva-
tion of a company for going international.

In the end, the company *Volkswagen* will be analyzed according to their individual
process of internationalization in order to integrate a practical example into the anal-
ysis. *Volkswagen* has been chosen due to the fact that this transnational enterprise,

compared to the other two big automotive concerns in Germany (BMW and Daimler-Benz), displays both an extended and quantitative and qualitative significant history of internationalization. (Pries: 1999)

## 2. How firms internationalize

### 2.1 The Product Life Cycle

In order to answer the question whether there is a best possible way for international-ization one has to take a look at sequential development paths a firm could pass through on its way of becoming an internationally operating firm.

The Product Life Cycle, henceforth referred to as PLC, expresses a traditional view on the internationalization process. The PLC takes over the assumption of the TNC having a strong domestic position before expanding geographically and this is why the PLC is a suitable notion to start with when observing the internationalization pro-cess of TNCs.

According to Dicken, Vernon starts from the premise that it is easier to introduce a new product in the own domestic market than it is from elsewhere. He further argues that this new product would as a result best reflect domestic characteristics. In the following, the PLC is explained in more detail by including the United States as an example. The PLC is divided into five different phases, which look as follows:

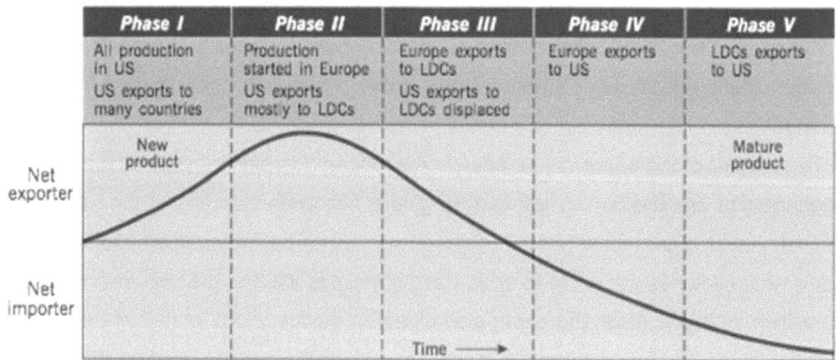

Figure 1: Dicken, Peter (2012): The product life cycle as an evolutionary sequence of US TNC's develop-ment. p. 117.

The first phase is characterized by domestic production and by the satisfaction of overseas demand by exports. In case of saturation of the domestic market and consistent profitability locally, the expansion into markets beyond the domestic market is indispensable. As a result of the distribution of economic activities, the US reduces its production and distribution costs and at the same time its market position can be improved.

The second phase is characterized by starting overseas production in other high-income countries. At this point, new production facilities abroad displace the US former exports and thus these exports can go to places where production has not begun yet. The hereby-saved production costs make exports to third country markets or even back to the US domestic market possible, which is displayed in phase three and four of the PLC.

In the fifth phase of the cycle, production is completely standardized and the US can eventually shift its production to low-cost locations in developing countries. (Dicken: 2012)

However, this approach has been criticized with respect to its lack of illustrating the more multifaceted and complex structure of TNCs. It furthermore leaves the more likely cross investment between the industrialized countries out of consideration. (Dicken: 2012) Others take the view that the PLC does not follow a set pattern but that it is rather the result of selected marketing strategies than their cause. The particular phases of the cycle are temporally not predictable because they differ from product to product. (Michel; Pifko: 2009) All these arguments show that it is necessary to take a look at different ways of TNC progress, which is done in the next passage of this paper.

## 2.2 Different ways of TNC progress

At first appearance, the representation of different ways of TNC progress can be classified as being very similar to the PLC outlined above. Throughout this section of the paper, however, it will become clear that this approach of TNCs progression is more open and leaves room for interpretation.

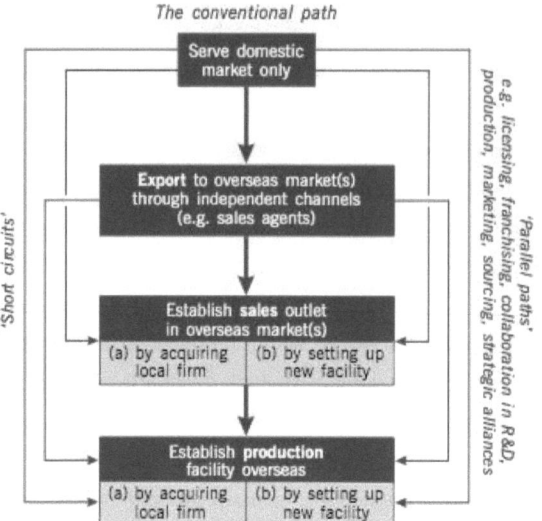

*The conventional path*

Figure 2: Dicken, Peter (2012): Diverse paths of TNC evolution. p. 118.

As within the PLC, this illustration shows that the firm is solely being a domestic firm in the beginning. It then appears that the company starts exporting into overseas markets through independent channels, which could be sales agents for instance. (Dicken: 2012) Usually a company starts its exporting activities as the domestic market suffers from saturation and therefore domestic sales decrease. The easiest way is direct exporting. Hereby, a company ships its goods after receiving payment. Subsequently, sales are likely to expand and this leads to a first sales office in the domestic market. After a while, sales offices are even set up in overseas markets. (Stonehouse [et al.]: 2004)

After establishing a sales outlet in the overseas market, a new production facility is located abroad, either by acquiring a local firm or by setting up a completely new facility. Even though, a firm is not necessarily restricted to this progression. It is possible to pass through a different development by consulting no intermediaries as the sales agents or by leaving out complete stages of the development. (Dicken: 2012)

To name but a few, other possibilities of running different paths are *licensing* or *franchising,* where an agreement is obtained which allows a partner to produce or sell abroad. Another opportunity provides the entering of a *joint venture* in which firms form an alliance and make equity investment. (Hamilton; Webster: 2012)

## 2.3 Born Global Firms

A *born global* firm can be defined in many ways. According to Cavusgil and Knight, Oviatt and McDougall are of the opinion that a born global firm is "a business organization that, from inception, seeks to derive significant competitive advantage from the use of resources and the sale of outputs in multiple countries" (Cavusgil; Knight: 2009, Intro.). Anderson and Evangelista even add a number to their definition and claim that a *born global* is a firm "in which foreign sales account for at least 25% of the total within three years of its inception [...]" (Anderson; Evangelista: 2006). *Born global* firms can thus be characterized by early internationalization since they intend to trade internationally right from the beginning or at least after a very short period of time. Their domestic country, which is usually of limited size, only serve to support their international business. Consequently, the size is one reason why those firms are forced to enter the global market very quickly and it also hints at the fact that *born global firms* mostly enter close partnerships with suppliers, distributors or consumers. (Hamilton; Webster: 2012)

This approach may challenge traditional views on the process of internationalization of a firm and there are certainly a lot more distinctive features of *born global firms* (see Cavusgil; Knight: 2009, p. 10) but in the course of this paper, it is necessary to at least show an interest in this business structure since *born global firms* play a major role in the discourse of internationalization.

## 3. Why firms internationalize

In nowadays business, international competitiveness is of high relevance. In this context, it is essential for a firm to be able to assert itself against foreign competitors without state subsidies or tariff protection. This holds true for both exporting and importing companies. The former need to become prevalent to international competitors on foreign markets, whereas the latter need to take up with competitors on the domestic market by means of import. A firm asserts oneself on the market when it has the ability to retain its market share and at the same time obtains the costs of production, including the opportunity costs of equity. (Sell: 2003)

Which markets are favored and what resources contribute to a firm's success is summarized in that which follows. It is of relevance that each of the factors mentioned in the following can take effect singly or in combination with each other. (Grünig; Morschett: 2012)

### 3.1 Market Access

Markets are not equally accessible. In the past, the major barrier for a firm to enter a specific market was transportation costs. Those costs are not totally unimportant today but their consideration has at least become less significant. Nowadays trade barriers, such as political constraints, play a more important role than charges for transport. (Dicken: 2012) Nevertheless, in the pre-1990s state protections were of higher degree than today, although emerging markets are still not stout today. (Sauvant: 2008) Attracted markets are often protected from imports, for instance by tariffs. That is to ascertain that a certain amount of the locally produced products are moreover sold locally. (Hamilton; Webster: 2012) Furthermore, legal or formal-institutional differences of the foreign market opposite to the domestic market can constitute serious barriers on behalf of a market entry. (Hess; Paesler: 2009) In case of free choice of the market, firms often decide on markets due to their size and structure of demand. Certainly, there are huge geographical variations when it comes to these economic factors, since there are enormous variations in terms of income and the demand for goods and services in a particular country. (Dicken: 2012) Increasingly popular markets are emerging markets in consequence of their relatively high rates of economic growth and purchasing power. Developed country markets, in comparison, grow only very slowly.

An enormous advantage of gaining market access on the whole is the approaching of costumers and sometimes even the possibility of gaining face-to-face contact. (Hamilton; Webster: 2012) This is particularly important when the domestic market is stagnating or even decreasing. Due to the approaching of new markets, the company has a high chance of gaining additional turnover and margin.

Another general benefit that occurs when entering a new market and going international is the balancing of risk. By emerging at different geographical markets, a company sometimes has the possibility of compensating lack of demand in one region. This is a factor, however, that seldom takes singly effect but rather occurs in combination with other drivers of internationalization. (Grünig; Morschett: 2012)

## 3.2 Lower Production Costs

One of the major goals of a company is to cut the costs of the production process. An important way of doing so is *offshoring*. It can be defined as the transferring of jobs abroad in order to economize high labor costs. The increasing costs of labor in rich countries force firms to look for cheaper locations in developing countries. This is especially done by labor-intensive industries, such as the textile, clothing or footwear industry. At first, low-level, unskilled or semi-skilled work was transferred to labor-cheap countries. Meanwhile, higher-skilled work or even polluting activities are being relocated and transferred. (Hamilton; Webster: 2012) Another important aim, which can be fulfilled due to the accessing of new markets, is the possibility of spreading fixed costs across more products so that economies of scale will emerge. In some industries, development expenses are high so that this is essential for remaining a substantial economic business. (Grünig; Morschett: 2012)

## 3.3 (Natural) Resources

The location near the source of natural resources plays a huge role particularly for natural resource industries. Their extractive activities are dependent on such re-sources and lay the foundation of natural-resource-orientated foreign investments. Not only natural resources but also generic resources, as for instance knowledge, skills, databases or even the relationships with other firms (*joint ventures*), act a part when it comes to a firm's prosperity. (Hamilton; Webster: 2012 and Dicken: 2012) In this context, the important role of the Internet is to mention, which boosted communi-cation and made the transfer and exchange of information easier, faster and cheap-er. If a firm is dependent on having knowledge on the spot, it is advantageous to lo-cate within a geographical cluster where the access to knowledge and technological know-how is very easy. (Dicken: 2012) TNCs have additional advantage over nation-al companies, since they are able to apply knowledge gained in one country to an-other country which they are also operating in. (Grünig; Morschett: 2012)

Another large geographical variation, which is also highly related to the choice of an appropriate market, can be found in educational levels or wage costs. Some indus-tries have just much higher levels of education and wage than others. (Dicken: 2012)

Up to this point in this section, the focus lay on the motivational factors of internation-alization. But in the context of transnational companies, the following paragraphs deal with the pressure of globalization and localization on companies, which, as a result, intentionally decide on becoming a TNC.

The drivers for a transnational strategy are both globally and locally orientated. The benefits of global embedding and local responsiveness need to be combined in order to successfully operate as TNC. (Stonehouse [et al.]: 2004)

### 3.4 Transnational Strategy

The following illustration shows the transnational strategy, which can be basically divided into global strategy options and local strategy options.

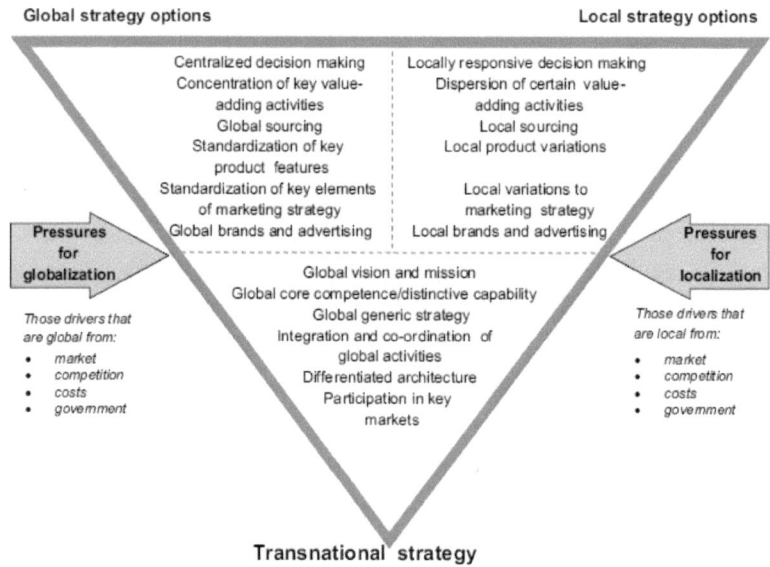

Figure 3: Stonehouse [et al.] (2004): Transnational Strategy. p. 190.

Global strategy options include a *global vision*, which is essential in order to take a global perspective on business activities. *Global knowledge-based core competences* are also of high importance so that a company can meet the challenges emerging on the global market. Transnational strategy is also uniquely defined over global *generic strategy* and *global coordination*, which signifies an application of global strategy based on core-competency as well as the global value-adding activities with

8

worldwide coordination. The *differentiated architecture* directly related to transnational strategy is reflected in the structuring of activities in order to maximize global advantages. Here, some activities are pooled while others are dispersed. A TNC *participates in key markets* while still regarding the entire world as potential market.

As yet, the activities mentioned are likely to be global. The following activities can be either global or localized constituents in transnational strategy due to transitions in the macro-environment of a company or in regard to the drivers of globalization. To these belong for instance *decision-making conditions* or *value-adding activities*, which may be both dispersed and concentrated or one of the two. Furthermore, *products, marketing strategy, branding* and *sourcing* belong to this category. The former may be standardized or adapted, whereas the global or localness of marketing strategy and branding are dependent on customer and product characteristics. For the latter, namely sourcing, location advantages decide on resources being globally or locally sourced. (Stonehouse [et al.]: 2004)

To sum up, the transnational strategy can be seen as a solution that combines the benefits of both global and local radius of orientation, whereas a global or more local orientation is dependent on the pressure for globalization or localization, which are the drivers that are local or global from market, competition, costs and governmental regulations that can be seen on the right-hand and left-hand side of the illustration above.


## 4. Case Study: *Volkswagen*

Due to the increasing international and interregional integration of economic activities, it is becoming more and more difficult for TNCs to secure regional competitive advantage and to decrease production costs within the PLC. (Kulic: 2009) Since many years, the automobile industry constitutes the most important branch of trade within the triad (Europe, Japan, NAFTA). This becomes particularly obvious when taking a look at some striking numbers: According to the *Organisation Internationale des Constructeurs d'Automobiles* (OICA), the annual automobile industry sales amounts to about 1.89 quintillions Euro. Accumulated, the automobile industry therefore embodies the sixth biggest economy of the world. This volume of sales is borne by more than eight million employees working in this industry. Furthermore, the automobile industry is the biggest driver of innovations worldwide. Research and development expenditures account for 70 billion Euros. The importance of Research and

development expenditures show that automobile products count as most technologi-
cal-intensive and dynamic products on nowadays markets. (Schömann: 2012)

In the following section of this paper, the company *Volkswagen* and its way of internationalizing and becoming a TNC will be analyzed. This is done by introducing
*Volkswagen's* distribution-orientated multinational company (MNC) phase, before
showing an interest for *Volkswagen's* production-orientated MNC phase. The last
step will deal with its final way of becoming a transnational company.

In comparison to the other German automobile companies, *Volkswagen* not only developed its global activities most but was also the first company to start internationalizing the production structures of its concern. This is particularly interrelated with
*Volkswagen's* product range, which was traditionally mass production-orientated and
restricted to middle and bottom market segment. (Pries: 2009) The name
*Volkswagen*, which can be translated with "People's Car" (Thompson; Martin: 2005),
already indicates its initial mass-production orientation. At an early opportunity after
National Socialism, *Volkswagen* evolved itself into a distribution-orientated MNC, on
the basis of the *VW-Käfer*, which was its regular product during that time. At this
stage, *Volkswagen* had the distinction of worldwide sales and production facilities in
many countries, whereas the American continent was the regional centre of sales
activities outside Germany. The production model across Volkswagen, the *Käfer*,
was soon also defining for overseas locations. (Pries: 2009) Under Nordhoff's guid-
ance as chief representative of *Volkswagen* in Wolfsburg (Becker: 2010), employees
and their lobby were embedded in a strategy of growth and spatial expansion. On the
basis of a relatively simple product for mass-demand and fordistic production structures (low degree of mechanization, simple production technologies, semi-skilled
qualification, etc.), this enabled *Volkswagen* both a materialistic and politic-social in-
volvement in the company's growth. (Pries: 2009)

This production model also shaped *Volkswagen's* internationalization strategy. In
1947, the first *Käfer* were exported to the Netherlands, and in 1949 also to the US.
During the 50s and 60s, production facilities were pre-eminently established on the
American continent. This was mostly done by initially founding sales companies, then
mounting automobiles out of imported CKD-sets (Completely-Knocked-Down-sets)
and finally developing complete production facilities. (Pries: 2009) CKD-sets impli-
cate that construction sets were completely mounted by a partner abroad. (Kutsch-

ker; Schmid: 2008) In order to make clear the extent to which the *VW-Käfer* was produced, the following table is included:

| Start: | Country: | Typ: | End: | Duration: |
|--------|----------|------|------|-----------|
| 1981 | Egypt | Assembly | n. a. | n. a. |
| 1954 | Australia | Assembly | 1975 | 21 |
| 1954 | Belgium | Assembly / Production | 1972 | 18 |
| 1955 | Brazil | Assembly / Production | 1986 | 31 |
| 1966 | Costa Rica | Assembly | 1975 | 9 |
| 1972 | Indonesia | Assembly | 1976 | 4 |
| 1951 | Ireland | Assembly | 1957 | 6 |
| 1973 | Yugoslavia | Assembly | 1976 | 3 |
| 1968 | Malaysia | Assembly | 1977 | 9 |
| 1954 | Mexico | Assembly / Production | 1998 | 44 |
| 1954 | New Zealand | Assembly | 1972 | 18 |
| 1975 | Nigeria | Assembly / Production | 1987 | 12 |
| 1966 | Peru | Assembly | 1987 | 21 |
| 1959 | Philippines | Assembly | 1982 | 23 |
| 1964 | Portugal | Assembly | 1976 | 12 |
| 1968 | Singapore | Assembly | 1974 | 6 |
| 1951 | South Africa | Assembly / Production | 1979 | 28 |
| 1972 | Thailand | Assembly | 1974 | 2 |
| 1961 | Uruguay | Assembly | 1987 | 26 |
| 1963 | Venezuela | Assembly | 1981 | 18 |

Figure 4: own illustration based on Pries, L. (1999): Auf dem Weg zu global operierenden Konzernen? BMW, Daimler-Benz und Volkswagen: Die *Drei Großen* der deutschen Automobilindustrie. München und Mering. p. 39.

Particularly in Brazil, Mexico and South Africa, highly integrated locations of production with remarkable mass production were developed in the 60s. As recently as the 70s, a fundamental change, and at the same time a transition to a product-orientated concern, took place. After a domestically boom until 1992, Volkswagen came to be a TNC. In regard to the concerns internationalization profile, this found expression in an enormous expansion of international activities in general (new facility in Changchun/China in 1991, Joint Venture with Ford in Palmela/Portugal starting 1991, cooperation and development of production facilities in the Czech Republic, Poland, Hungary, Taiwan, Malaysia, Indonesia and the Philippines). Second, this change found expression in a qualitative shift of functions between countries and locations. The production facility in Puebla/Mexico serves as an appropriate example to underline this statement. After closing the Westmoreland facility in the US in 1988, the supply of the US-market with the *Golf/Vento A2* was attributed to the Mexican site. Due to several different reasons (problems with partial delivery, qualitative problems, changes in the exchange-rate, etc.), the market share of *Volkswagen* in the USA further declined. (Pries: 1999) To ensure completeness it is incidentally stated that the market share in the US additionally increasingly decreased and in the mid-2000 it only accounted for two percent. In the mid-2000s, the market share of *Volkswagen* in the USA amounted only two percent. This was due to insufficient adapting to the demands of US-customers. The lack of simple things, as cup holders for instance, resulted in the decline of the market share to a minimum. The weak ability of *Volkswagen* to adapt to the US-market is also revealed in the late market entry of US-customized models like the *SUV* or the *Minivan*. (Grünig; Morschett: 2012)

A serious industrial struggle in 1992 and the therefore extensive changes in management, labor relations and especially in labor organization marked the point of culmination in the prevailing production model. Against this background, a new role was attributed to the facility in Mexico. For the first time a non-European facility took worldwide responsibility for the production run of a new product, in this case for the *New Beetle*. Innovations did not even necessarily come from the parent plant in Germany anymore, which was concerned with the production model across *Volkswagen* before.

Additionally, the targeted pursuit of the platform strategy was an essential characteristic of the transition of *Volkswagen* becoming a TNC. A consequent implementation of this product- and production philosophy exerted high pressure towards a standardization of market strategies across *Volkswagen*. It allowed regionalization of products and sales strategies with simultaneous centralization of research and development for platforms across *Volkswagen*. This strategy also benefited the homogenization of quality standards and the adaption of technical production level. It furthermore simplified systematic benchmarking and production flexibility between locations, which in turn intensified intragroup competition concerning both products and product quotas. The platform strategy was part of a qualitative accelerating process of globalization and the modernization of the concern, which sets the phase of becoming a TNC apart from the previous periods of the internationalization of the concern. (Pries: 1999)

Nowadays, the TNC *Volkswagen* is located at 94 production sites and it is therefore embedded in a worldwide production network. Sales in 153 countries enable the concern to employ 501,956 employees worldwide. Within its everyday production of about 34,500 vehicles of all kind, the company successfully uses the ability of linking knowledge from all its sites. By now, the company also deals successfully with both customer's demands and market requirements throughout the world. *Volkswagen's* ability to flexibly adapt to those factors has increased. (Volkswagen: 2012) In the course of globalization, *Volkswagen* (as well as other automobile companies) needs to deal with and adapt to the increasing complexity, which is revealed in increasing distances of transport and resulting replacement time but also costlier planning through additional relations and intermediaries. (Göpfert; Braun; Schulz: 2012)

## 5. Conclusion

The aim of this paper was to answer the following questions:

1. Is there a best possible way for internationalization?

2. What economic incentives play a role when a firm decides to go international?

Thus, the present paper applied to the motivational factors of transnationalization and moreover to the process of internationalization itself.

One possibility to systematically illustrate the process of internationalization was found by means of the PLC. However, the cycle is accompanied by certain restrictions and serves pre-eminently to illustrate the process of internationalization traditionally. In this context, different paths of TNC progress were demonstrated. Hereby it became obvious that companies have various possibilities of going international and that they are often forced to accept detours in order to operate on the more and more globalized and complex market. The option of internationalizing was contrasted with *born global firms*. It became clear that there are also companies that are global due to their structure and their need to operate internationally right from the beginning.

On closer consideration it can be stated that there is no definite process a company passes through on its way of becoming international. In fact, the portraying of different paths of internationalization already indicates that each company needs to decide on its own process according to its individual structure and its own objectives. The process therefore is influenced by various factors, which were pointed out in the second part of this paper. Further on, it was given priority to the three main motivational factor of going international: market access, lower production costs and access to resources. The combination of all these factors helps a company to successfully hold its ground in global competition, which was regarded as most important element in nowadays business. In order to use the mentioned factors in a sufficient way, companies make use of the benefits of global orientation, whereas transnational companies extend this global strategy by at the same time maintaining local responsiveness on their domestic market. The numbers revealed in the current status of Volkswagen show how *Volkswagen* successfully implemented the most important elements in its internationalization process. Therefore, the concern highly successful withstands exterior challenges and increasing competition on the globalized market with respect to local conditions.

# 6. Bibliography

## 6.1 Research Literature

ANDERSON S. and F. EVANGELISTA (2006): The Entrepreneurship in the Born Global Firm in Australia and Sweden. In: Journal of Small Business and Enterprise Development. Vol 13, Issue 4, p. 642.

BECKER, H. (2010): Darwin's Gesetz in der Automobilindustrie. Warum deutsche Hersteller zu den Gewinnern zählen. Heidelberg.

CAVUSGIL, S.T. and G. KNIGHT (2009): Born Global Firms. A New International Enterprise. New York.

DICKEN, P. (2012). Global Shift. Mapping the Changing Contours of the World Economy. Sixth edition. New York. London.

GILLIES, G.L. (2012): Transnational Corporations and International Production Concepts, Theories and Effects. Second edition.

GÖPFERT, I., D. BRAUN and M. SCHULZ (2012): Automobillogistik. Stand und Zukunftstrends. Wiesbaden.

GRÜNIG, R. and D. MORSCHETT (2012): Developing International Strategies. Going and Being International for Medium-sized Companies. London. New York.

HAMILTON, L. and P. WEBSTER (2012): The International Business Environment. Second edition. New York.

HESS, M. and PAESLER, R. (2009): Wirtschaft und Raum. Wege und Erträge der Münchner wirtschaftsgeographischen Forschung. Band 20. München.

KUTSCHKER, M. and S. SCHMID (2008): Internationales Management. Sixth Edition. München.

LULE, Jack (2012): Globalization and Media. Global Village of Babel. Maryland.

MICHEL, S. and C. PIFKO (2009): Marketingkonzept. Grundlagen mit zahlreichen Beispielen, Repititionsfragen mit Lösungen und Glossar. Zürich.

PRIES, L. (1999): Auf dem Weg zu global operierenden Konzernen? BMW, Daimler-Benz und Volkswagen: Die Drei Großen der deutschen Automobilindustrie. München und Mering.

SCHÖNMANN, S.O. (2012): Produktentwicklung in der Automobilindustrie. Managementkonzepte vor dem Hintergrund gewandelter Herausforderungen. Wiesbaden.

SELL (2003): Einführung in die internationalen Wirtschaftsbeziehungen. Second edition. München.

STONEHOUSE, G., J. HAMILL, D. CAMPBELL and T. PURDIE (2004): Global And Transnational Business. Strategy and Management. Second Edition. West Sussex.

THOMPSON, J. and F. MARTIN (1995): Strategic Management. Awareness and Change. Fifth edition. London.

KULIC, D. (2009): Automobilindustrie zwischen Globalisierung und Regionalisierung. Ist der Freihandel nur eine Illusion? Hamburg.

15

## 6.2 Online Sources

VOLKSWAGEN (2012): Volkswagen International. URL:
http://de.volkswagen.com/de/unternehmen/laenderauswahl.html (Accessed: 29.07.12)

## 6.3 List of Figures

**Figure 1:** DICKEN, P. (2012): Global Shift. Mapping the Changing Contours of the World Economy. Sixth edition. New York. London. p. 117.

**Figure 2:** DICKEN, P. (2012): Global Shift. Mapping the Changing Contours of the World Economy. Sixth edition. New York. London. p. 118.

**Figure 3:** STONEHOUSE [et al.] (2004): Transnational Strategy. p. 190.

**Figure 3:** own illustration based on PRIES, L. (1999): Auf dem Weg zu global operierenden Konzernen? BMW, Daimler-Benz und Volkswagen: Die Drei Großen der deutschen Automobilindustrie. München und Mering. p. 39.